Technical English 1

Workbook

Christopher Jacques

Pearson Education Limited

Edinburgh Gate
Harlow
Essex CM20 2JE
England

and Associated Companies throughout the world.

www.pearsonelt.com

First published 2008

Fifteenth impression 2019

ISBN: 978-1-4058-9647-4 (book)

ISBN: 978-1-4058-9653-5 (book for pack)

Set in Adobe Type Library fonts

Printed in Slovakia by Neografia

Acknowledgements

The publishers and author would like to thank the following for their invaluable feedback, comments and suggestions, all of which played an important part in the development of the course: Eleanor Kenny (College of the North Atlantic, Qatar), Julian Collinson, Daniel Zeytoun Millie and Terry Sutcliffe (all from the Higher Colleges of Technology, UAE), Dr Saleh Al-Busaidi (Sultan Qaboos University, Oman), Francis McNeice, (IFOROP, France), Michaela Müller (Germany), Małgorzata Ossowska-Neumann (Gdynia Maritime University, Poland), Gordon Kite (British Council, Italy), Wolfgang Ridder (VHS der Stadt Bielefeld, Germany), Stella Jehanno (Centre d'Etude des Langues/ Centre de Formation Supérieure d'Apprentis, Chambre de Commerce et d'Industrie de l'Indre, France) and Nick Jones (Germany).

Illustrated by Mark Duffin, Peter Harper and HL Studios

The publisher would like to thank the following for their kind permission to reproduce their photographs:

(Key: b-bottom; c-centre; l-left; r-right; t-top)

Alamy Images: CrashPA 8 (1); alveyandtowers.com: 8 (4); Andrew Holt Photography: 16bc; Art Directors and TRIP photo Library: 16t, 19; aviation-images.com: Mark Wagner 8 (3); Britain on View: Lee Mawdsley 38; Corbis: Bo Zaunders 8 (6); Charles O'Rear 42c; David Kimber: 9 (middle far left), 9 (middle left), 9 (middle right), 9 (middle far right), 9 (bottom far left), 9 (bottom left), 9 (bottom right), 9 (bottom far right), 9tl, 9tc, 9tr; DK Images: 16tl, 16r, 16bl; Eye Ubiquitous / Hutchison: Bob Battersby 52; Freeplay Energy: 15; Joel Brown www.syntheticimage.net: Joel Brown www.syntheticimage.net 8 (2); Los Alamos National Laboratory: 63; NASA: 40; PA Photos: 29; Michel Spingler 42b; PunchStock: Corbis 48 (4); Punchstock / Uppercut: 48 (1); Rex Features: Andy Lauwers 8 (5); Goran Algard 50; Image Source 48 (2), 48 (3); Robert Down: Robert Down / photographersdirect.com 32; Science Photo Library Ltd: TRL Ltd 28; Travel Library Ltd, The: Stuart Black 42t; Woods Hole Oceanographic Institution: 58

Cover image: *Front*: iStock Photo: Kristian Stensoenes

All other images © Pearson Education

Picture Research by: Kevin Brown

Every effort has been made to trace the copyright holders and we apologise in advance for any unintentional omissions. We would be pleased to insert the appropriate acknowledgement in any subsequent edition of this publication.

Designed by Keith Shaw

Cover design by Designers Collective

Project Managed by David Riley

Contents

1 Basics

1 Use the words in the box to complete the dialogues.

> what's   where   what   I'm   is   are   I'm

1 A: Hi, _____ Kaito.
   B: Hello, my name _____ Pedro.
   A: Nice to meet you.
2 A: Hello. _____ are you from?
   B: I'm from Japan. _____ is your name, please?
   A: I'm Hans. Pleased to meet you.
3 A: Good to meet you, Svetlana. _____ you from Poland?
   B: No, _____ from Russia. _____ your name?
   A: I'm Danielle.

2 Use the words in the pool to complete the orders.

1 Stand _____.
2 Write _____.
3 Turn _____.
4 Close _____.
5 Sit _____.
6 Raise _____.
7 Come _____.

> right
> down
> your name
> your book
> in
> up
> your hand

3 Write the words in the correct columns.

> adapter   antenna   bolts   cable   chisel   nuts   plug   saw   screwdriver
> screws   spanner   washers

Tools	Electricals	Fixings
_____	*adapter*	_____
_____	_____	_____
_____	_____	_____
_____	_____	_____

2 Letters and numbers

1 ▶💿 **02** Listen and correct the five mistakes on the business card.

2 ▶💿 **03** Listen and complete the form.

Surname:	J _____
First name:	_____
Company:	_____
Email address:	_____

3 Match items 1–10 with the right words. Then match items 11–20.

1	gal		a)	amp
2	€		b)	angle/degree
3	kg		c)	Celsius
4	A		d)	euro
5	in		e)	foot
6	ft		f)	gallon	*1*
7	km		g)	gram
8	°		h)	inch
9	g		i)	kilogram
10	C		j)	kilometre

11	+		k)	kilometres per hour
12	m		l)	kilowatt
13	kW		m)	litre
14	V		n)	metre
15	kph		o)	negative
16	rpm		p)	positive	*11*
17	W		q)	pound
18	L		r)	revolutions per minute
19	£		s)	volt
20	–		t)	watt

4 ▶ 🔘 04 Mr Martin is buying a car. Listen and write down the facts about the car.

1   Kilometres: *120 000* km
2   Engine temperature: _____ ° Celsius
3   Petrol tank: _____ litres
4   Engine speed: up to _____ rpm
5   Top speed: _____ kph
6   Price: _____ euros

3 Dates and times

1   Write the words for these ordinal numbers.

4th	*fourth*	5th	_____
12th	_____	29th	_____
23rd	_____	8th	_____
7th	_____	31st	_____
30th	_____	6th	_____
22nd	_____	20th	_____

2   Complete the puzzles.

1   Jan 31 Fri → Feb 8

January the thirty-first is a Friday, so February the eighth is a Saturday.

2   Mar 29 Wed → Apr 2

_____

3   May 29 Tue → June 3

_____

4   July 30 Thur → Aug 4

_____

5   Sept 28 Mon → Oct 7

_____

6   Nov 27 Thur → Dec 6

_____

3   Use the words in the box to complete the dialogue.

that's   is   it's   then   what   when's   it's

A: _____ the meeting?
B: _____ on Monday.
A: _____ that Monday 12th?
B: Yes. _____ right.
A: Do you know _____ time?
B: _____ at 10 o'clock.
A: OK. See you _____. Bye.
B: Bye.

4 Word list

NOUNS	NOUNS	ORDINAL NUMBERS	VERBS
adapter	amp	first	listen
antenna	angle	second	lower
bolt	Celsius	third	pick up
cable	degree	fourth	put down
chisel	euro	fifth	raise
nut	foot	sixth	read
plug	gallon	seventh	say
saw	gram	eighth	sit
screw	inch	ninth	stand
screwdriver	kilogram	tenth	start
spanner	kilometre	eleventh	stop
washer	kilometres per hour	twelfth	write
counter	kilowatt	thirteenth	**ADVERBS**
flight	litre	twentieth	closed
model	metre	thirtieth	down
platform	pound	**PHRASES**	in
first name	revolutions per minute	Excuse me	left
surname	volt	Hello	off
initial(s)	watt	Good to meet you	on
	ADJECTIVES	Nice to meet you	open
	negative	Pleased to meet you	out
	positive		right
			up

1   Make up answers to these questions. Use words from column 2 of the Word list.

1   How heavy is it? *425 grams*   *22 kilograms*

2   How hot is it? _____

3   How long is it? _____ _____ _____

4   How far is it to Dubai? _____

5   How fast is the car travelling? _____

6   How fast is the engine turning? _____

7   How much petrol is in the tank? _____ _____

8   What's the price of the car? _____ _____

9   How do you write *225 V* in words?

_____

1 Naming

1 Write sentences for the pictures.

Parts	Vehicles
axle deck nose number plate tail wheel	boat motorbike mountain bike plane racing car rocket

1 _That's the wheel of a racing car._

2 _____

3 _____

4 _____

5 _____

6 _____

2 Use the words in the box to correct the sentences.

| bolts   nails   nuts   screw   screwdriver   spanner   staple   washers |

1 _That isn't_ **a hammer. That's** _a screwdriver._

2 _Those aren't_ **screws. Those** _are nails._

3 _This_ _____ **a chisel. This** _____.

4 _____ **washers. These** _____.

5 _____ **a nail. This** _____.

6 _____ **nuts. These** _____.

7 _____ **a staple. That's** _____.

8 _____ **nuts. Those** _____.

2 Assembling

1 How do you change a car wheel? You need:

a **jack**, to raise and
lower the car

a **box spanner**,
for the nuts

a **spare wheel**

Complete the instructions for the pictures, using the verbs from the box.

loosen   lower   put on   raise   take off   tighten

1 _____ the car with the jack.
2 _____ all the nuts with the box spanner.
3 _____ all the nuts.
4 _____ the wheel _____ the axle.
5 _____ the spare wheel _____ the axle.
6 _____ all the nuts.
7 _____ all the nuts with the box spanner.
8 _____ the car.

2 Write the dialogue lines in the right order.

30 mil. How many nails do you need?	Shopkeeper: _____
30 mil, please.	Customer: _____
Hello.	Shopkeeper: _____
I need 80, please.	Customer: _____
Some nails. What size do you need?	Shopkeeper: _____
Hello. I need some nails, please.	Customer: _____

3 Ordering

1 ▶ 🌐 **05** Listen to the two phone messages. Correct the mistakes in the names and numbers.

| 1 Name: Vladislaw Sczetin | Phone number: 00 48 920 4516 |
| 2 Name: Abdel Mohamed Mabruk | Phone number: 00 20 537 1490 |

2 ▶ 🌐 **06** Listen to the two phone messages. Complete the message forms.

1

Date: _____

Time: _____

Caller: _____

Phone number: _____

2

Date: _____

Time: _____

Caller: _____

Phone number: _____

3 ▶ 🌐 **07** Listen to the dialogue. A customer is ordering skateboard parts on the phone. Complete the order form.

SKATEBOARDERS
ORDER

Surname: _____

Address: _____

_____

Postcode: _____

Tel: _____

Item (circle)	Colour (circle)			Size (circle)			Quantity (write)
Helmet	red	yellow	blue	large	medium	small	_____
Deck	red	yellow	blue	large	medium	small	_____
Pad	red	yellow	blue	large	medium	small	_____

4 Word list

NOUNS	NOUNS	VERBS	ADJECTIVES
axle	bolt	assemble	large
deck	hammer	loosen	medium
helmet	lever	pull	small
nose	nail	push	red
pad	nut	put	yellow
plate	screw	take	blue
tail	screwdriver	tighten	
truck	spanner	use	
wheel	staple		
	washer		

1 Spelling: there are eight words in the Word list with double letters. Write them here.

wheel, _____

_____

2 Vocabulary groups: write the words in column 2 on the correct line.

Tools: *hammer,* _____

Things: *bolt,* _____

3 Complete the instructions for skateboarding with words from the box.

loosen	push	put	take	tighten

Before skateboarding

_____ on the helmet.

_____ it down onto your head.

_____ the helmet strap.

_____ on the pads.

_____ the pads.

After skateboarding

_____ the pads and _____ them off.

_____ the helmet strap and _____ off the helmet.

Strap

Section 1

1 Complete the dialogues.

| I'm | he's | that's | is | do | I'm | are |

1 A: _____ you Maria?
   B: No, _____ Sonia. _____ Maria.
2 A: What _____ you do, Toni?
   B: _____ a builder.
3 A: _____ Carlos a builder?
   B: No, _____ an electrician.

2 Check the information in Students' Book page 9. Write the dates in column 2.

A person writes ...	What is the date?
1 Claire Paris, 1/2/11	February 1st 2011
2 Vicky Chicago, 3/9/11	_____
3 Yuki Tokyo, 11/01/22	_____
4 Matt Seattle, 11/12/11	_____
5 Director, ISO Geneva, 2011.07.08	_____
6 Peter Berlin, 9/10/11	_____

3 Work out the sequence of days and dates. Write the missing ones.

1 Monday, May the first
2 *Thursday, May the fourth*
3 Sunday, May the seventh
4 _____
5 Saturday, May the _____
6 _____
7 Friday, May the _____
8 _____

Section 2

1 Jumbled letters. Write the plural words.

1 lotsb   *bolts*   5 ilsan   n_____

2 hessraw   w_____   6 lesax   a_____

3 wressc   s_____   7 eatsksarbod s_____

4 tuns   n_____

2 Write two more dialogues, like the example. Use the words from the box.

A: What's this tool called?

B: It's a *spanner*.

A: Is it for *nails*?

B: No. It's for *nuts*.

hammer   screws   nuts   spanner   screwdriver   nails

A: What's this tool called?

B: It's _____

A: Is _____

B: _____

A: _____

B: _____

A: _____

B: _____

3 Complete the dialogue with the questions.

What's your phone number?   What's your email address?

What's your name?   What size cards do you need?

How many do you need?   What's your address and postal code?

When do you want them?

A: Hello. I need to order some business cards.

B: *How many* _____

A: 200, please.

B: _____

A: 85 millimetres by 55 millimetres.

B: _____

A: Stevens, with a V. Initials HC.

B: _____

A: 14 Hayfield Road, Bristol. BR7 4JK

B: _____

A: 0117 893462.

B: _____

A: It's harry.stevens@ojs.com

B: _____

A: Friday, please.

1 Tools

1 Complete the crossword. Find a twelfth word in the puzzle.

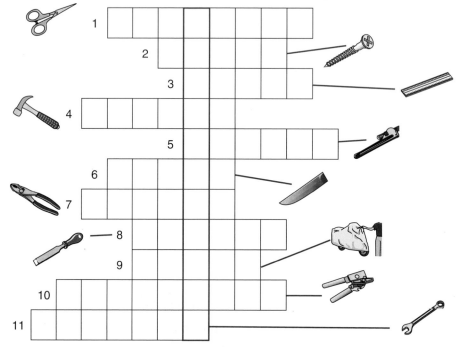

2 Write the answers to the puzzles. Use each item once.

| hammer | pair of pliers | pair of scissors | saw | screwdriver | spanner |

1   It has a handle, a shaft and a head. It turns screws. It is a *screwdriver*.
2   It has a shaft and a head. It drives in nails. It is a _____.
3   It has two handles and two blades. It cuts paper. It is a _____.
4   It has a shaft and jaws, but no blades. It tightens nuts. It is a
    _____.
5   It has two handles, jaws and blades. It cuts wire. It is a _____.
6   It has a handle and a blade. It cuts wood. It is a _____.

3 Use the words in the box to complete the dialogues.

| do | does | don't | doesn't | have | has |

1   A: _____ Carlos need a spanner?
    B: No, he _____.
    A: _____ he need a pair of pliers?
    B: Yes, he _____.
    A: Does he _____ a saw?
    B: Yes, he _____ two.

2   A: _____ you have a hammer?
    B: No, I _____
    A: _____ you need a hammer?
    B: Yes, I _____.
    A: I don't _____ one. Go and ask
    Pedro. He _____ one in his
    tool box.

2 Functions

1 Match the word halves and write the words next to the explanations.

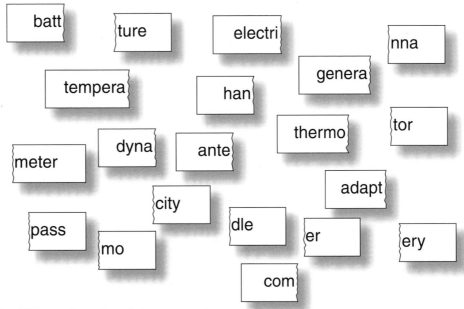

1 This makes electricity. *generator*

2 This shows North. _____

3 This stores electricity. _____

4 An AC _____ changes AC to DC.

5 This receives radio signals. _____

6 A solar panel changes sunlight into _____ .

7 You can measure _____ in Fahrenheit or Celsius.

8 You turn this round with your hand. _____

9 This measures temperature. _____

10 This turns and makes electricity. _____

2 Use the verbs from the box to complete the text.

charge	shine	charges	turn	listen	turns	produces

Are you going on holiday? This 3-in-1 torch, radio and battery charger is for you.

When you (1)_____ the handle, it (2)_____ the dynamo. This (3)_____ the battery. You can then (4)_____ the torch, or (5)_____ to the radio.

For example, five minutes at 120 rpm (6)_____ enough power to listen to the radio for twenty minutes. You can also turn the handle to (7)_____ your mobile phone.

3 Locations

1 ▶💿 **08** Listen to the dialogue in the factory. Where does the driver put the boxes?

1 speakers	2 keyboards	3 DVD players
4 scanners	5 headphones	6 amplifiers
7 mouse pads	8 adapters	9 printers

1   Listen and write the product number on the right shelf.

2   Write all the product names on the right shelves.

3   Look at the shelves. What order are the products in?

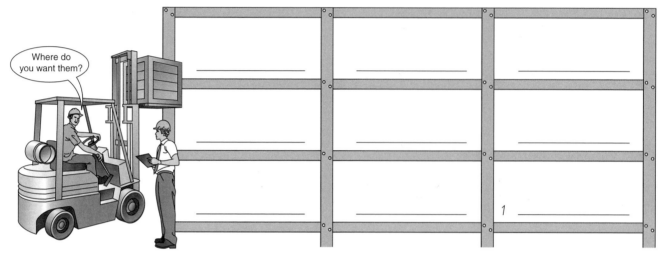

2 ▶💿 **09** Listen to a dialogue on a boat. Where do the people put the things? Write the number of the location (1–12) next to the word on the right.

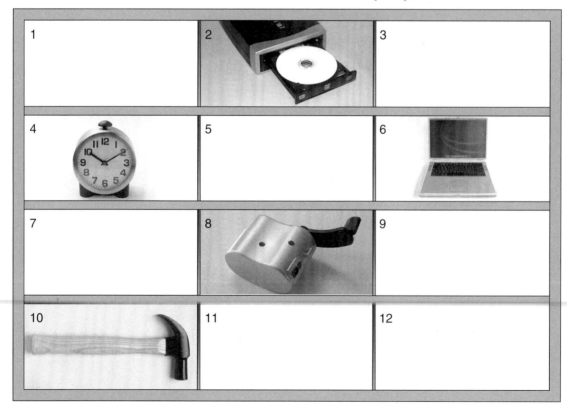

multi-tool
pliers
radio
wrench
batteries
torch
scissors *12*

4 Word list

NOUNS (tools)	NOUNS (electricity)	VERBS	ADJECTIVES
blade	alarm	change	external
boat	battery	charge	internal
bottle opener	clock	connect	plastic
building site	dynamo	cut	**PHRASES**
can opener	electricity supply	drive in	at the bottom
compass	generator	grip	at the top
cover	mains electricity	measure	in the centre
handle	radio	produce	in the middle
head	solar panel	receive	on the left
jaws	solar power	shine	on the right
key tool	torch	turn	above
metal	**NOUNS (computer)**		below
multi-tool	computer		to the left of
pick	computer station		to the right of
pliers	cursor		
ruler	DVD drive		
scissors	keyboard		
shaft	mouse		
string	printer		
survival tool	scanner		
thermometer	screen		
wire	speaker		
wrench			

1 Match each noun in column 1 with a phrase in column 2.

1 Chisels   a) loosen screws.

2 Hammers   b) tighten nuts.

3 Pliers   c) cut wood.

4 Rulers   d) drive in nails.

5 Saws   e) cut metal.

6 Scissors   f) grip wire.

7 Screwdrivers   g) measure everything.

8 Wrenches   h) cut paper.

4 Movement

1 Directions

1 Look at the pictures of the jump jet.

1 Which picture shows a vertical take-off? (Picture _____)

2 Which picture shows a short take-off? (Picture _____)

3 Which directions can you see? Write the letters from the pictures (A–D) here.

vertically up _____ horizontal _____ diagonally up _____

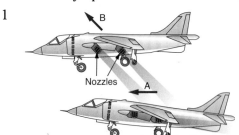 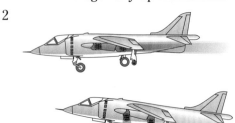

2 Which directions can the jump jet fly? Complete the text with words from the box.

> forwards   sideways   straight down   straight up   to the right   up and down

The jump jet can fly like a helicopter or fly like a passenger plane. The jump jet has one engine and four nozzles. The four nozzles can point straight down. Then the jet engine lifts the plane (1)_____ into the air. In the air, the four nozzles can rotate and point backwards. This pushes the plane (2)_____. Then the plane can fly at about 1165 kph. Like a passenger plane, it can turn to the left or turn (3)_____. It can fly diagonally (4) _____. It can also fly backwards and (5)_____, a little. How does it land? It stops in the air and flies (6)_____.

3 Read about the movements of the human leg. Complete the text with words from the box.

> angles   ankle   degrees   directions   hip   knee   move   pivots   rotate
> sideways

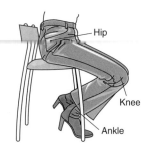

The leg has three (1) *pivots*, the hip, the knee and the ankle. The ankle can move in three (2)_____. At the (3)_____, the foot can move up and down about 50 (4)_____. It can (5)_____ from side to side about 50 degrees, and it can (6)_____ about 15 degrees. The (7)_____ can move in the same directions, but with different (8)_____. The (9) _____ can only move in one direction. At the knee, the lower leg can only move up and down. It cannot move (10)_____ or rotate.

2 Instructions

1 ▶ 🔵 **10** Write the full forms. Then listen and check.

1 30 kph   *thirty kilometres per hour*
2 500 rpm _____
3 15 m/s _____
4 65 mph _____
5 8 km/s _____

2 ▶ 🔵 **11** Listen and write the speeds. Use the short forms from question 1.

1 Sound travels at _____.
2 The engine of a Formula 1 car turns at about _____.
3 The moon truck Apollo 16 Rover travels at _____.
4 A solar-powered car can travel at _____.
5 A person on skis can go downhill at _____.
6 A person on a snowboard can go downhill at _____.
7 The maximum speed of a train in France is _____.
8 The fastest sailing ship sails at _____.
9 A Blackbird jet flies at _____.

3 ▶ 🔵 **12** Listen to the dialogue. Are all the parts for the radio-controlled truck in the box? Listen and tick the things on the list.

Instruction manual

Transmitter

Truck

Antenna for transmitter

Antenna for truck

2 9V batteries

4 Use the words from the box to complete the text about the truck.

| control | moves | press | receives | sends | turns | use |

The transmitter (1)_____ radio signals to the receiver in the truck. An antenna on the truck (2)_____ signals from the transmitter. The truck and the transmitter (3)_____ electricity from batteries. Six buttons (4)_____ the speed and direction: forwards, backwards, forward and left, forward and right, backwards and left, backwards and right. There are two electric motors. One motor (5)_____ the wheels to the left or right. The other motor drives the back wheels forwards or backwards. (6)_____ the control button 'Forwards'. The motor turns the shaft and the shaft turns the axle. The truck (7)_____ forward.

3 Actions

1 Read the instruction manual. Write the letters (A–H) from the diagram next to the names of the controls.

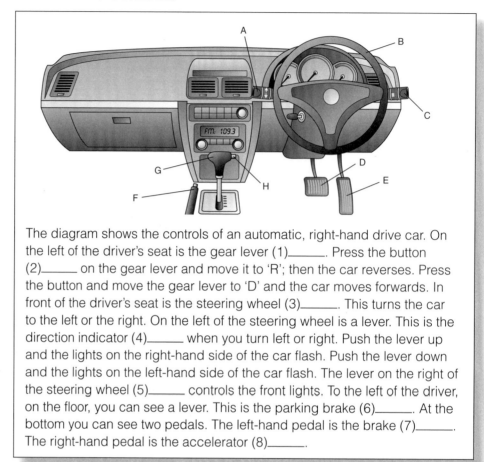

The diagram shows the controls of an automatic, right-hand drive car. On the left of the driver's seat is the gear lever (1)_____. Press the button (2)_____ on the gear lever and move it to 'R'; then the car reverses. Press the button and move the gear lever to 'D' and the car moves forwards. In front of the driver's seat is the steering wheel (3)_____. This turns the car to the left or the right. On the left of the steering wheel is a lever. This is the direction indicator (4)_____ when you turn left or right. Push the lever up and the lights on the right-hand side of the car flash. Push the lever down and the lights on the left-hand side of the car flash. The lever on the right of the steering wheel (5)_____ controls the front lights. To the left of the driver, on the floor, you can see a lever. This is the parking brake (6)_____. At the bottom you can see two pedals. The left-hand pedal is the brake (7)_____. The right-hand pedal is the accelerator (8)_____.

2 Write instructions for driving a car. Write full sentences from these notes. Use *when* and *you*, and add *the* and punctuation.

1 pull gear lever to 'R' → car reverses

   When you pull the gear lever to 'R', the car reverses.

2 pull gear lever to 'D' → car moves forwards

   _____

3 press accelerator → car goes faster

   _____

4 press brake pedal a little → car goes slower

   _____

5 turn steering wheel to the right → car turns right

   _____

6 turn steering wheel to the left → car turns left

   _____

7 press brake pedal → car stops

   _____

3 Put the instructions for parking a car in the correct order. Complete the instructions with the following words: *forwards, left, right*.

Order: _____

A Drive _____ a little and turn the steering wheel to the _____.

B Reverse a little more and turn the steering wheel to the _____. Stop.

C Drive _____ slowly. Stop.

D Reverse and turn the steering wheel to the _____.

4 Word list

NOUNS (tools)	NOUNS (electricity)	VERBS	ADJECTIVES
accelerator	parking brake	accelerate	backwards
angle	pedal	ascend	down
antenna	pivot	control	forwards
brake	plane	descend	sideways
direction	revolution	dock	up
elbow	robot	park	to the left
forearm	roll	press	to the right
handle	shoulder	pull	**PHRASES**
helicopter	slider	push	horizontal axis
joystick	speed	reverse	vertical axis
kilometre	steering wheel	rotate	
lever	switch	slide	
metre	tilt	slow down	
mile	wrist	turn round	

1 Find nine nouns for driving a car. Write them here.

_accelerator_____

_____

2 Find opposites in columns 3 and 4 for the following words and write them here.

accelerate _____

ascend _____

pull _____

forwards _____

up _____

to the left _____

3 Find seven verbs for flying a helicopter. Write them here.

Helicopters can accelerate, _____

_____

Section 1

1 Look at the diagram of the work station. Tick the true statements. Correct the false ones.

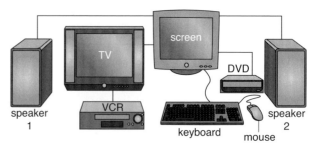

1   The screen is in the centre. ✓
2   The keyboard is in the centre, ~~above~~ the screen. *below*
3   The TV is to the right of the screen.
4   The VCR is on the left, below the TV.
5   Speaker 1 is on the right.
6   Speaker 2 is on the left.
7   The mouse is at the top, to the left of the keyboard.
8   The DVD drive is below the mouse, to the left of the screen.

2 Where are the programmes on the screen? Make sentences with the words in the box.

> bikes   cars   football   the news   boats   science   skateboards   space
> planes

1   *Football is at the top, on the left.*
2   *Planes are at the top, in the centre.*
3   _____
4   *Bikes are on the middle line,*
      _____
5   _____
6   _____
7   _____
8   _____
9   _____

3 Write the singular form of the words in the box. If a word has no singular form, write 'a pair of …'.

> batteries   hammers   overalls   pincers   pliers   scissors   spanners   wrenches

1   Singular form: *battery* _____ _____
      _____

2   No singular form: *a pair of overalls* _____ _____
      _____

Section 2

1 Find letters in the diagram (A–D) for each sentence. Use the phrases from the box to complete the sentences.

> descend   up and down
> forwards and backwards
> rotate   diagonal or horizontal

1 (   ) The crane can move _____ on its wheels.

2 (   ) The top part of the crane can _____ through 360°.

3 (   ) The arm of the crane can ascend and _____ through 90°. It can be in a vertical, _____ position.

4 (   ) The hook below the end of the arm can go _____.

2 Use the words in the box to complete these questions and answers. Then match the questions with their answers.

> is   are   do   does   can   can't   put   need   press   goes   receives

1 _____ you find the user manual?

2 How _____ the truck work?

3 Where _____ I put the battery?

4 Where _____ the antenna go?

5 How _____ I steer the truck?

6 _____ there two batteries in the box?

7 _____ we need a second battery?

a) No, there _____ only one.

b) You _____ it in the transmitter.

c) No, I _____ find it.

d) Yes, we _____ it for the truck.

e) It _____ on top of the truck.

f) It _____ signals from the transmitter.

g) You _____ one of the control buttons.

3 Read the instructions (A–D) for steering a boat backwards. Put them in the correct order. Then complete the instructions with the following words: *forwards, left, centre, backwards*.

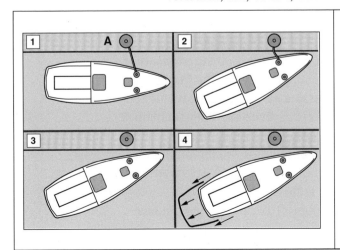

1 1 _____   2 _____   3 _____   4 _____

A Turn the steering wheel to the _____ position. Pull the lever _____; this puts the engine into reverse. Reverse slowly.

B Turn the steering wheel to the left. Push the engine lever forwards; this moves the boat slowly _____ and to the _____.

C Pull the engine lever to the _____ position. Loosen the rope. Take off the rope from Point A.

D Start the engine. Tie the rope on the _____ of the boat to Point A.

5 | Flow

1 Heating system

1 Draw a line from each word to its opposite.

sink   above   bottom   out of   cold   cool   enter   outlet   push

hot   inlet   leave   heat   pull   rise   top   below   into

2 Rewrite the sentences. Change the words in italics. Use words with opposite meanings from question 1.

1   A solar panel *heats* water. A fridge … → *A fridge cools water.*

2   *Hot* water *rises* to the *top* of a water tank. →
_____

3   The *inlet* pipe for *cold* water is *below* the pump. →
_____

4   Water *enters* the tank through the *inlet* pipe. →
_____

5   *Push* the shower head *into* the pipe. →
_____

3 Look at the diagram. Warm water comes up from underground and heats water for the houses. Use the verbs and prepositions in the box to complete the description of the heating system.

flow	leave	push	rise	above
below	into	through	to	out of

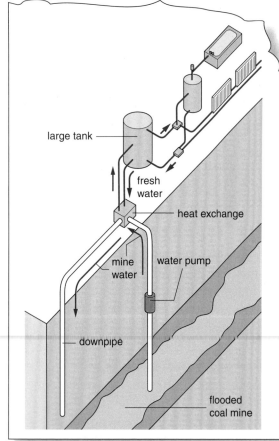

large tank

fresh water

heat exchange

mine water

water pump

downpipe

flooded coal mine

In this system, there are houses (1) *above* a flooded coal mine. At 170 metres (2)_____ ground, the temperature of the mine water stays at 14.5 °C. The water pump brings up the mine water and (3)_____ it (4)_____ the heat exchanger. The mine water comes (5)_____ the heat exchanger and (6)_____ back into the coal mine (7)_____ the downpipe.

In the heat exchanger, the temperature of the fresh water (8)_____ to 55 °C. This warm water then flows to a large tank. Then it (9)_____ the large tank and goes (10)_____ the houses.

2 Electrical circuit

1 Match the words in the box to sentences 1–7.

> battery   cable   controller   lamp   solar panel   electrical current   switch

1 shines a light when the switch is on: *lamp*
2 converts the sun's energy into an electrical current: _____
3 stores electricity: _____
4 When a _____ is closed, the electrical current can flow.
5 DC is a type of _____.
6 Electricity passes through the _____ to the lamp or the battery.
7 carries the electrical current: _____

2 ▶ 🔊 **13** Look at the diagram for a water-wheel and a generator which supplies current to a workshop next to the river. Complete the sentences with the present simple. Then listen and check your answers.

1 If the river is high, and the workshop is open, *the current flows from the generator into the workshop.* (current / flow / generator / workshop)
2 If the river is high, and the workshop is closed, _____ _____. (current / flow / generator / batteries)
3 If the river is low, and the workshop is open, _____ _____. (current / flow / batteries / workshop)
4 If the river is low, and the workshop is closed, _____ _____. (current / not / flow)
5 If the batteries are full, _____ _____. (current / not / flow / generator / batteries)
6 If the batteries are empty, _____ _____. (current / not / flow / batteries / workshop)

3 ▶ 🔊 **14** Listen to the dialogue. Circle the correct specifications for the items.

1 Solar panels   a) 4 × 16 W   b) 40 × 60 W   c) 4 × 60 W
2 Controller   a) 1 × 3 A   b) 1 × 5 A   c) 1 × 15 A
3 Batteries   a) 4 × 12 V, 50 Ah   b) 4 × 12 V, 100 Ah   c) 4 × 15 V, 150 Ah
4 Lamps   a) 6 × 20 V, 8 W   b) 16 × 12 V, 18 W   c) 6 × 12 V, 8 W
5 Cable   a) 2.5 mm, 30 amps   b) 6 mm, 53 amps   c) 16 mm, 100 amps
  (12 metres)

3 Cooling system

1 Complete these sentences for a world weather forecast. Write the temperatures as words.

1 The night-time temperature in Helsinki will be *minus two degrees Fahrenheit*. (–2 °F)

2 The day-time temperature in Mexico City will be *twenty-one degrees Celsius*. (21 °C)

3 The day-time temperature in Los Angeles will be _____. (75 °F)

4 The coldest night-time temperature in Moscow will be _____. (–8 °C)

5 The day-time temperature in Tunis will be _____. (24 °C)

6 The highest day-time temperature in Karachi will be _____. (33 °C)

2 Use the words in the box to answer the questions with short answers. Use some of the words twice.

cool water	engine	fan	hot water	thermostat	two hoses	water pump

1 What pushes cool water round the engine? *The water pump*

2 What connects the radiator to the engine? _____

3 What controls the temperature of the engine? _____

4 What flows from the engine to the radiator? _____

5 What blows air through the radiator? _____

6 What sinks to the bottom of the radiator? _____

7 What cools the water in the radiator? _____

8 What passes along the bottom hose and back to the engine?

_____

9 What drives the water pump? _____

3 Look at the diagram for a watering system. Complete the sentences with the words in the box. Put the verbs into the present simple.

around	at the top	at the bottom	from	into	out of	through

1 From the spring, water (flow) *flows* to a reservoir *at the top* of the hill.

2 _____ the reservoir, water (pass) _____ _____ a pipe to the field.

3 The pipe (go) _____ _____ a field of fruit trees.

4 Water (leave) _____ the pipe _____ small holes.

5 The water then (flow) _____ _____ the fruit trees.

6 A little water (flow) _____ _____ the bottom of the field.

7 This water (enter) _____ a tank _____ of the hill.

4 Word list

HEATING AND COOLING		PREPOSITIONS OF MOVEMENT	ELECTRICAL
NOUNS	**VERBS**		**NOUNS**
engine	blow	around	battery
fan	connect	into	cable
hose	control	out of	conductor
inlet	cool	through	controller
radiator	drive	to	electrical circuit
shower head	enter		electrical current
solar panel	flow		energy
thermostat	go		lamp
valve	heat		solar panel
water pipe	leave		switch
water pump	move		**VERBS**
water tank	pass		convert
	push		flow
	rise		shine
	sink		short-circuit

1 Complete the sentences with verbs from column 2.

1 Cold water _____ the system through the inlet.
2 Water _____ into the tank through a pipe.
3 The sun _____ the water in the solar panel.
4 Hot water _____ to the top of the tank.
5 Cold water _____ to the bottom of the tank.
6 Hot water _____ the system through the shower head.

2 Match the sentence halves.

1 The water pump pushes   a) the temperature of the water.
2 The thermostat controls   b) air through the radiator.
3 The two hoses connect   c) the hot water from the engine.
4 The fan blows   d) water around the engine.
5 The radiator cools   e) the radiator to the engine.

6 | Materials

1 Materials testing

1 Make sentences about the materials with 'can ..., but ... can't', or 'can ... and ... can'.

1 (bend / metal / wood) *You can bend metal, but you can't bend wood.*

2 (heat / air / water) *You can heat air and you can heat water.*

3 (melt / plastic / wood)
_____

4 (scratch / glass / metal)
_____

5 (stretch / nylon / glass)
_____

6 (break / glass / wood)
_____

7 (cut / wood / metal)
_____

8 (compress / air / glass)
_____

2 A lecturer is showing a DVD of a test. Complete the description. Use the present continuous.

Hello. Now we can watch the DVD of a car crash. Here they (1)*are testing* (test) the material for the seatbelt. The human dummy (2)_____ (sit) in the test car. This dummy weighs 90 kilos. Here the technician (3)_____ (tighten) the nylon seatbelt around the dummy. Now the technician (4) _____ (start) the engine of the radio-controlled car.
Look at the crash in slo-mo (= slow motion). The car (5)_____ (run) into the concrete block at 40 kph. The body of the dummy (6)_____ (stretch) the nylon seatbelt. And see, the dummy (7)_____ (touch) the airbag.
Look carefully. (8)___ the dummy's face _____ (strike) the front window?
No, it isn't. There is no contact with the front window.

3 Write questions and answers for the pictures.

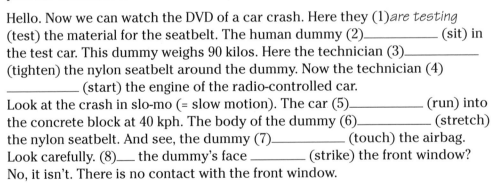

1 you / push / handles ?
2 he / walk ?
3 she / bend / wall bars ?
4 you / pull / bar / down ?
5 he / push / bar ?
6 she / bend / legs ?

1 A: *Are you pushing the handles?*   4 A:
   B: *No, I'm rowing.*   B:

2 A: 5 A:
   B: B:

3 A: 6 A:
   B: B:

2 Properties

1 Find the names of 14 materials in the puzzle and circle them. The words go vertically from top to bottom, and sideways from left to right. No words go diagonally.

B	A	J	L	O	Y	C	O	M	P	O	S	I	T	E
P	L	A	S	T	I	C	E	T	O	Z	P	R	A	K
L	U	R	T	I	B	K	Y	L	L	B	O	J	L	I
O	M	A	L	J	M	O	Q	A	Y	U	L	S	D	A
F	I	B	R	E	G	L	A	S	S	I	Y	T	I	Y
B	N	S	D	R	A	R	X	P	T	B	C	N	A	O
T	I	T	A	N	I	U	M	D	Y	F	A	H	M	I
J	U	E	K	Y	L	B	N	T	R	I	R	V	O	Z
A	M	E	B	L	C	B	F	G	E	A	B	H	N	I
J	R	L	K	O	Q	E	S	V	N	U	O	Z	D	W
Y	Z	C	O	N	C	R	E	T	E	X	N	B	G	Y
H	I	R	J	T	K	U	L	C	E	R	A	M	I	C
S	V	N	X	P	G	R	A	P	H	I	T	E	Q	W
I	Y	B	T	L	E	K	O	E	U	J	E	C	D	I

2 Underline the two correct adjectives for each material.

1   A ceramic cup is flexible/<u>heat-resistant</u> and <u>hard</u>/soft.
2   A concrete floor is rigid/flexible and brittle/tough.
3   A rubber tyre is rigid/flexible and weak/strong.
4   A fibreglass window frame is heat-resistant/soft and rigid/flexible.
5   A nylon rope is rigid/flexible and strong/weak.
6   The graphite in the middle of a pencil is light/heavy and hard/soft.
7   A polycarbonate road sign is rigid/flexible and strong/weak.
8   A polystyrene coffee cup is brittle/tough and heavy/light.

3 Design a plane. Choose one material for each part of the plane.

1   (nose cone / plastic / aluminium)
   The nose cone is made of aluminium.

2   (wheels / fibreglass / aluminium alloy)
   _____

3   (tyres / ceramic / rubber composite)
   _____

4   (frame / composite / polystyrene)
   _____

5   (inside / fibreglass / rubber composite)
   _____

6   (seats / plastic / ceramic)
   _____

7   (engine / fibreglass / aluminium alloy)
   _____

8   (wings / aluminium alloy / plastic)
   _____

3 Buying

1 ▶ 🎧 **15** Listen and complete the order form. A customer is buying equipment on the phone.

THE CLIMBING SHOP	
ORDER FORM	
Date: *23/03/08*	**Helmet** (polycarbonate / fibreglass) (L / M / S)
Product name: _____	
Product no: _____	**Rope** (nylon / nylon + rubber composite) (50 m / 75 m / 100 m)
Quantity: _____	
Colour: _____	**Jacket** (cotton / polyester) (XL / L / M / S)
Size: _____	
Material: _____	**Backpack** (nylon / polyester) (XL / L / M / S)
Price: _____	

2 ▶ 🎧 **16** Listen and correct the email addresses.

1 jclark@eyeway.co.uk → _____
2 alex2@antigm.ac.uk → _____
3 s.hagen@renault.fra → _____

3 ▶ 🎧 **17** Listen and write the website addresses.

1 News: _____
2 Live radio: _____
3 Radio-controlled toys: _____

4 A customer is phoning a sports shop. Write questions for the answers.

1 Q: *What's your surname, please?*
   A: It's Badrawi.
2 Q: _____
   A: B–A–D–R–A–W–I.
3 Q: _____
   A: 01273 497 633.
4 Q: _____
   A: Ali dot badrawi at atlas dot com.
5 Q: _____
   A: Yes. A–L–I dot badrawi at atlas, that's A–T–L–A–S dot com.
6 Q: _____
   A: I need three helmets.
7 Q: _____
   A: I'd like white ones, please.
8 Q: _____
   A: I want to pay in euros, please.

4 Word list

NOUNS (Materials)	NOUNS (Car parts, other)	VERBS	ADJECTIVES
alloy	backpack	bend	brittle
aluminium	cone	break	corrosion-resistant
ceramic	engine	burn	flexible
composite	frame	climb	hard
concrete	helmet	coat	heat-resistant
cromoly	jacket	compress	heavy
diamond	piston	corrode	light
fibreglass	radiator	drop	rigid
graphite	rope	heat	soft
nylon	spoiler	hold	strong
plastic	tyre	melt	tough
polycarbonate	vehicle	row	weak
polyester	wheel	run	**PHRASES FOR EMAILS**
polystyrene	wing	scratch	
rubber		stretch	dash
steel		strike	dot
titanium		touch	forward slash
			hyphen
			underscore

1 Memory test. What is a racing car made of? Write the materials from column 1.

1 The nose cone *is made of fibreglass*.

2 The wheels *are made of* _____

_____.

3 The frame _____.

4 The tyres _____.

5 The radiator _____.

6 The engine _____.

7 The pistons are coated with _____.

8 The wings are made of _____ and

_____.

2 Write the opposites of the adjectives from the list in column 4.

1 Nylon isn't weak. It's *strong*.

2 Polystyrene isn't tough. It's _____.

3 Graphite isn't hard. It's _____.

4 Rubber isn't rigid. It's _____.

5 Aluminium isn't heavy. It's _____.

Review

Section 1

1 Use the words from the box to complete the phone dialogues.

about	are	here	here	how	I'm	OK	thanks	that	this

1 A: Hello?

B: Hello. Is (1)_____ Paulo?

A: Yes.

B: It's Sven (2)_____.

A: Oh, hi, Sven.

B: Hi. How (3)_____ things?

A: Great, (4)_____. How are you?

B: I'm (5)_____.

2 A: Hello. Mona Hall (6)_____.

B: Oh, hi, Mona. (7)_____ is Ingrid.

A: Hi, Ingrid.

B: Hi. (8)_____ are you?

A: Very well. How (9)_____ you?

B: (10)_____ fine, thanks.

2 Write the -ing forms of the verbs on the correct line.

bend	climb	cut	dive	drive	drop	grip	heat	hold
leave	move	pull	push	rise	run	sit	strike	swim

1 Add -ing: *bending*, _____

_____

2 Double the last letter and add -ing: *cutting*, _____

_____

3 Drop the -e and add -ing: *diving*, _____

_____

3 Complete the dialogue about the engine's cooling system. Put the verbs into the present continuous. One verb is used twice.

blow	drop	go	push	rise	run	work

A: Is everything OK?

B: No. The engine's cooling system *isn't working*. The temperature of the water _____.

A: _____ the fan _____ air through the radiator?

B: Yes, the fan is fine.

A: _____ the pump _____ water round the engine?

B: Yes, the pump is working.

A: Look! That clip on the bottom hose is loose. Water _____ out of the hose. So the cold water _____ not _____ back to the engine. Tighten the clip.

B: _____ the water _____ out of the hose now?

A: No. Check the temperature.

B: Ah! The temperature _____. Good!

Section 2

1 Match phrases from the table to make sentences.

warm ice cubes	sink
pull a rubber band	burn
strike a ceramic cup very hard	break
heat water to 100 °Celsius	stretch
cool water	melt
heat pieces of wood	boil

If you warm ice cubes, they melt.

_____

_____

_____

_____

_____

2 Read the text and complete the table below.

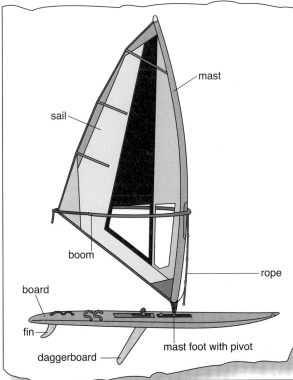

This sailboard is made from light, strong and flexible materials. The board is strong but light. It is made of polystyrene, coated with fibreglass. The mast is strong and flexible. It is made of polycarbonate. The mast and the boom support the sail. The boom is rigid and strong. It is made of aluminium, coated with rubber. The sail is light but strong. It is made of a mixture of nylon and polyester. Fixed to the end of the boom is a strong rope. It is made of nylon. The rigid daggerboard and fin are made of polycarbonate. There is a pivot at the foot of the mast. This is strong and flexible. It is made of rubber.

Part	Material	Properties
board	*polystyrene, fibreglass*	*strong, light*
mast		
boom		
sail		
rope		
daggerboard		
fin		
pivot		

Specifications

1 Dimensions

1 Use the words in the box to label the picture.

| bridge | cable | deck | pier | pylon | road | span | tunnel |

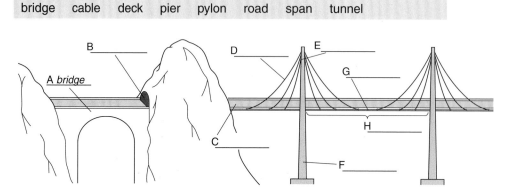

2 Make sentences. Write the words in the correct order.

1 270 has metres The sea of depth a → *The sea has a depth of 270 metres.*

2 deep is metres 25 river The → _____

3 is metres 330 span long The → _____

4 a 160 The height metres of have pylons → _____
_____

5 the 22 kilometres length The road of is → _____
_____

6 width has 8 deck The metres of a → _____

3 Write questions and answers about the bridge.

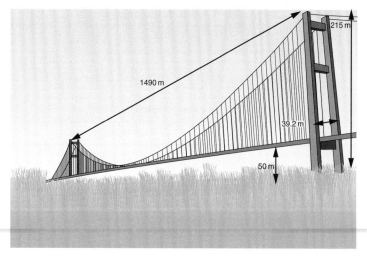

Runyang Bridge, China

1 where / this bridge ?
Q: *Where is this bridge?*
A: *It's in China.*

2 long / inner span ?
Q: _____
A: _____

3 high / pylons ?
Q: _____
A: _____

4 wide / deck ?
Q: _____
A: _____

5 high / deck / above water ?
Q: _____
A: _____

2 Quantities

1 Complete the text with facts from the specification chart.

The Gherkin

Address: 30 St Mary Axe, London
Completion date: 2003
Height: 180 m
Floors: 40
Glass area: 24 000 sq m
Floor area: circle
Footprint: small
Number of lifts: 18
Speed of lifts: 6 m/s
Materials: reinforced concrete,
    steel, aluminium, glass
Width of glass lens: 2.4 metres

This building is called 'The Gherkin'. It was completed in (1)_____. The
(2)_____-storey building is 180 (3)_____ high. The building is made of
(4)_____, (5)_____, (6)_____ and (7)_____. The glass windows
have an (8)_____ of 24 000 (9)_____. The building has (10)_____ lifts.
Each lift travels at 6 (11)_____. Each floor area of the building is a
(12)_____. The floors at the top and bottom are small. The floors in the
middle of the building are bigger. The footprint of the building is
(13)_____. All the glass on the side of the building is flat. But on the top of
the building, there is one round glass lens. It is 2.4 metres (14)_____.

2 🔊 **18** A customer is ordering some materials. Listen and complete the order form.

Item	Kind (circle)	Size (circle)	Product number (circle)	Quantity (write)
Paint	green / grey	5 / 10 litre tin	P176GR / D186G	
Cement	grey / white	10 / 20 kg bag	C0116W / S0196G	
Nails	packet of 50 / 100	24 / 30 mm	N420 / N240	
Screws	packet of 50 / 100	20 / 24 mm	S00941 / F00921	

3 🔊 **19** Listen and complete the questions from question 2.

1 *How much paint* do you need?

2 _____ _____ _____ do you need?

3 _____ _____ _____ do you need?

4 _____ _____ _____ do you need?

5 _____ _____ _____ do you need?

6 Do you have _____ _____?

7 _____ _____ _____ do you need?

8 Do you need _____ _____?

3 Future projects

1 Read the article about a new train tunnel. Complete the specification chart.

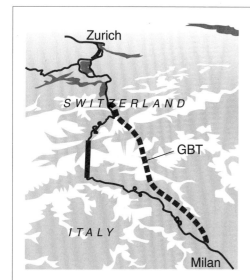

Zurich

SWITZERLAND

GBT

ITALY

Milan

The Gotthard Base Tunnel

The Gotthard Base Tunnel (GBT) will be the longest railway tunnel in the world. Engineers are building it now, in Switzerland, under the Alps. They will finish the project in 2016–2017.

Today, many trucks use the roads over the Alps. In future, they will use the GBT. The tunnel will connect Zurich in Switzerland with Milan in Italy.

There will be two tunnels. Each tunnel will be about 57 km long. They will run about 600 metres below the old St Gotthard railway tunnel (completed in 1881).

The new trains will use electricity. Some trains will carry trucks and cars. Fast passenger trains will travel at 250 kph. About 200–250 trains per day will use the GBT.

Gotthard Base Tunnel (GBT)	
Location of tunnel	*in Switzerland, under the Alps*
Possible completion date	
Number of tunnels	
Length of tunnels	
Depth below old tunnel	
Maximum speed of trains	
Source of power for trains	
Number of trains per day	

2 Change the long forms to short forms.

1 They are building a new tunnel. *They're building a new tunnel.*

2 There will be two new tunnels. _____

3 They will finish the tunnel in 2017. _____

4 The trains will not use magnetism. _____

5 There will be more than 200 trains per day. _____

3 Underline seven mistakes in this report. Then rewrite the report.

> The GBT will be the longest <u>road</u> tunnel in the world. It will connect Italy and France. Engineers will finish the project in 2011. The new tunnel will be above the old railway tunnel. There will be over 300 trains per day. The new trains will use diesel. All of them will run at 250 kph.

The GBT will be the longest railway tunnel _____

_____

_____

_____

4 Word list

NOUNS (Bridge)	NOUNS (Design)	NOUNS (Materials)	VERBS
cable	completion	aluminium	attach
deck	depth	cement	build
pier	design	fibreglass	fix
pylon	elevator	glass	lay
span	footprint	glue	make
NOUNS (Tunnel)	foundation	oil	put
compressed air	height	paint	**ADJECTIVES**
diesel	length	reinforced concrete	amazing
magnetism	location	steel	approximate
vacuum	material	superglue	deep
	quantity	**UNIT NOUNS**	high
	specification	bag	inner
	storey	bottle	long
	structure	packet	outer
	width	tin	super-fast
		tube	wide

1 Find nouns for these adjectives.

long – l_____   high – h_____   wide – w_____   deep – d_____

2 Make phrases from the words in the box. Write them below.

a	bottle tube bag packet tin	of	cement oil paint glue / superglue screws

_____   _____

_____   _____

_____

3 Choose a verb from the Word list and complete these phrases.

1 *lay* the foundations   5 _____ the deck

2 _____ the piers   6 _____ the deck to the cables

3 _____ the pylons on the piers   7 _____ the road

4 _____ the cables to the pylons

8 Reporting

1 Recent incidents

1 Complete the table of verb forms.

		Verb	Past simple	Past participle
Type A Regular	add -ed	check		
	add -d			changed
Type B Regular	double the final letter and add -ed		stopped	
		plan		
Type C Irregular	verb = past simple = past participle	cut		
			put	
Type D Irregular	past simple = past participle		bought	
				sold
			sent	
Type E Irregular	past simple ≠ past participle		fell	
		speak		
				taken
		write		

2 Complete the dialogue. A is the manager. B is the manager's assistant.

A: *Have you spoken* to Security?

B: Yes, I have.

A: Good. _____ the new customer?

B: No, I haven't. I'll do it now.

A: _____ an email to HTB?

B: Yes, _____.

A: Good. _____ the incident report?

B: No, _____. I'll do it now.

> THINGS TO DO
> speak to Security ✓
> ring the new customer
> send an email to HTB ✓
> write incident report

3 Yesterday, the police received a lot of phone calls. Complete the sentences with verbs and nouns from the box.

have stolen has crashed has broken have come has taken have jumped has driven have run	diamonds digger motor boat shop river town centre shop window sledgehammer window

1   Hello? Police? A thief *has taken* my *digger*.

2   Police? A man _____ a digger into the _____ here.

3   Hello? A digger _____ into a

_____ in Broad Street.

4 Help! Two men _____ into my
   _____ in Broad Street.

5 One man _____ the display case with a
   _____.

6 The two thieves _____ some
   _____.

7 Two men with bags _____ down to the
   _____.

8 The two men _____ into a
   _____. They are on the river now.

2 Damage and loss

1 🔊 **20** In each sentence, fill in the gap and underline the best verb. Then listen, check and repeat.

1 They _have_ <u>bent</u> / burnt the router antenna.
2 The user manual _is_ <u>torn</u> / burnt.
3 Someone _____ bent / broken the camera.
4 The body of the radio _____ cracked / cut.
5 The speakers _____ damaged / torn.
6 Someone _____ cut / bent the power cable.
7 The lenses of the goggles _____ cut / scratched.
8 I _____ burnt / broken my overalls.
9 They _____ dented / torn the car door.

2 🔊 **21** Listen to Part 1 of the dialogue and correct the customer details. Then listen to Part 2 and complete the damage report.

Order No:	PC08/1020/0017	Item	Damaged	Missing
Name:	Mr Bert Sandle	1 router antenna	_bent_	
Address:	14 Hayford Road	2 mouse		✓
	Catford	3 computer screen		
	London	4 keyboard		
Postcode:	SE10 4QU	5 power cable		
Tel:	0208 411 4009	6 LH speaker		
Email:	bsandell87@bdg.co.uk	7 RH speaker		
		8 user manual		

3 Complete the sentences. Some of the phrases are used more than once.

> are   doesn't have   has   is   there's   there are

Reporting damage	Reporting something missing
1 The box *is* damaged.	1 The headphones _____ missing.
2 The overalls _____ torn.	2 _____ no pliers in the toolbox.
3 _____ a dent on one of the speakers.	3 The power cable _____ a plug.
4 _____ some cracks on the body of the radio.	4 _____ no batteries in the box.
	5 The radio _____ no antenna.
	6 _____ no user manual in the box.

3 Past events

1 Read the diary of a space tourist. Then complete the interview below.

1 Q: which year / travel / to / ISS?
   Which year did you travel to the ISS?
   A: In 2008.

2 Q: when / you / take off ?
   _____
   A: On April 12th.

3 Q: how / travel / into space ?
   _____
   A: On the Space Shuttle.

4 Q: what / take / with you ?
   _____
   A: Six Luka cameras and my laptop.

5 Q: what / do / on / ISS ?
   _____
   A: I tested all the Luka cameras.

6 Q: you / repair / solar panel ?
   _____
   A: No. John repaired it.

7 Q: when / you / leave / ISS ?
   _____
   A: On April 20th.

8 Q: when / you / land / in / USA ?
   _____
   A: On April 21st.

> 12.04.08 Took off on Space Shuttle. Took 6 Luka cameras and laptop.
> 13.04.08 Shuttle docked with International Space Station (ISS)
> 14.04.08 Tested all 6 Luka cameras. All worked OK.
> 15.04.08 John did spacewalk. He repaired solar panel on ISS.
> 20.04.08 Left ISS, after 7 days.
> 21.04.08 Returned to earth. Landed in USA.

2 Ben damaged his laptop a month ago. He rang the IT hotline. Write his answers to the questions. Use the past simple + *ago*.

1 When did you buy your laptop? (10 months)
   I bought it 10 months ago.

2 When did you drop it? (4 weeks)
   _____

3 When did you phone the company? (3 weeks)
   _____

4 When did you bring it into the Service Department? (10 days)
   _____

5 When did you send your email? (3 days)
   _____

6   When did you receive our bill? (2 days)

_____

7   When did you ring? (10 minutes)

_____

4 Word list

VERBS (Irregular)	VERBS (Regular)	VERBS (Damage)	NOUNS (Building)
bend / *bent*	check	bend	beam
break	climb	break	brick
burn	crack	burn	bucket
buy	crash	crack	builder
cut	dent	cut	crane
drive	happen	dent	digger
fall	land	scratch	hard hat
fly	launch	tear	scaffolding
go	lift	**NOUNS (General)**	sledgehammer
lose	move	accident	**NOUNS (Space)**
put	order	ambulance	global navigation
sell	raise	body (of radio)	moon
send	repair	damage	satellite
speak	scratch	display screen	shuttle
steal	snorkel	fuse	space station
take	**PHRASAL VERBS**	goggles (plural)	space tourist
tear	break into (irregular)	insulation	space walk
write	pick up (regular)	lens	telescope
	put on (irregular)	overalls (plural)	
	take off (irregular)	spark plug	
		surface	

1   Write the past tenses of all the irregular verbs on the Word list. (See the example at the top of column 1.)

2   Complete the sentences with nouns from the Word list.

   1   They put the i_____ around the water pipe.

   2   He climbed up to the top of the s_____.

   3   He broke the bricks with a s_____.

   4   He put a h_____ h_____ on his head and started work.

   5   The c_____ lifted the metal beam onto the building.

   6   One of the builders drove the d_____ into a brick wall.

   7   He had an accident. He didn't put on his g_____ and damaged his eyes.

   8   He stopped work, took off his dirty o_____ and went home.

Section 1

1 Use the words in the box to complete the texts. Some words are used more than once.

| at | below | deep | depth | length | long | more than | through | wide | width |

The Corinth Canal

The Corinth Canal is in Greece. It is 6.8 kilometres
(1)_____ and 21 metres (2)_____. The
(3)_____ of the water in the canal is 8 metres.
Large ships cannot sail (4)_____ the canal, but
small tourist ships can. (5)_____ 11 000 ships
travel through the canal every year.

The TauTona Gold Mine

The TauTona gold mine is in South Africa. It has a maximum
(6)_____ of 3.5 km. The total (7)_____ of tunnels is
(8)_____ 800 km. Mine workers get to the bottom of the
mine in super-fast lifts. These travel (9)_____ 16 m/s.
The mine opened in 1957. Soon they will open a new mine. This will be 3.9 km
(10)_____.

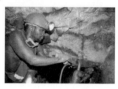

The Channel Tunnel

The Channel Tunnel is a railway tunnel between England
and France. It has two tunnels for trains. Each tunnel is
51.5 km (11)_____ and 7.7 metres (12)_____.
There is a small third tunnel for engineers. This has a
(13)_____ of 4.8 metres. The under-sea part of the
tunnel has a (14)_____ of 39 km. Most of the tunnel is 45 metres (15)_____
the sea floor. The travel time (16)_____ the tunnel is 20 minutes. Tunnels were
started in 1881 and in 1922. The present tunnel was completed in 1994.

2 Use the words in the box to complete the dialogues.

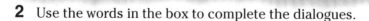

| any | how | many | many | much | one | six | some | some |

1 A: Hello. Can I help you?

   B: Yes. I'm building a wall and I need _____ cement.

   A: _____ much do you need?

   B: I need two bags please. And I also need _____ sand.

   A: How _____ bags do you need?

   B: I need _____ bags, please.

2 A: Hello. What can I do for you?

   B: Do you have _____ paint?

   A: Yes. How _____ do you need?

   B: 10 litres, please. And I need some nails.

   A: How _____ packets?

   B: _____ packet, please.

Section 2

1 Complete the table.

Verb	Past simple	Past participle
bend	bent	bent
build		
burn		
find		
lose		
pay		
break	broke	broken
come		
give		
go		

2 Ask questions about these recent incidents. Use the words in brackets.

1 A digger has driven into a shop window. (when / the / the)
When did the digger drive into the shop window?

2 Some thieves have broken into the office. (when / the)
_____

3 A mechanic has found some money. (how much / the)
_____

4 Some builders have taken off the old roof. (when / the)
_____

5 Some scaffolding has fallen down. (where / the)
_____

3 Complete the dialogue. A is the supervisor. B is the builder.

A: *Have you put up the scaffolding?*
B: *Yes, I have.*
A: *Good. Have you changed the power cable?*
B: *No, I haven't. I'll do it next week.*
A: _____
B: _____
A: _____
B: _____
A: _____

THINGS TO DO
put up scaffolding ✓
change the power cable ✗
buy the bricks ✗
speak to the supplier ✓.
order the water tank ✗

1 Operation

1 Use the words in the box to complete the text.

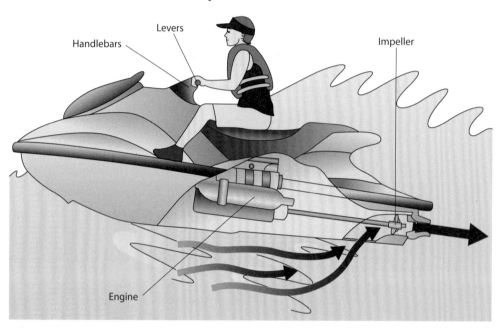

controls	drives	propels	pulls	pushes	release	steers	supports

Personal watercraft

A personal watercraft has a fibreglass body and an engine. A seat is mounted on the body and (1) *supports* the rider. The rider (2)_____ the craft with handlebars and (3)_____ the speed with levers. The engine is mounted on the body of the craft and (4)_____ the impeller. The impeller (5)_____ water in and (6)_____ it out. This (7)_____ the craft forwards. If you want to stop the craft, you (8)_____ the lever. This stops the engine.

2 Complete the questions and answers.

1 What *does* the engine do? It _____.
2 What _____ the impeller do? It _____.
3 What _____ the handlebars do? They _____.
4 What _____ the levers do? _____
5 What _____ the seat _____? _____

3 Make sentences with the words. Use 'mounted on' or 'attached to'.

1 seat / body *The seat is mounted on the body.*
2 handlebars / body _____
3 levers / handlebars _____
4 engine / body _____

2 Hotline

1 Find these things in the picture in question 2. Write the words below.

| adapter | disk drive | display | mouse | power switch | speaker |

1 *display*   3 _____   5 _____
2 _____   4 _____   6 _____

2 🔊 22 Listen to three short dialogues between a customer and a service technician. Parts of some of the dialogues don't match the picture. Listen and circle your answers.

Dialogue 1: no mistakes / 1 mistake / 2 mistakes

Dialogue 2: no mistakes / 1 mistake / 2 mistakes

Dialogue 3: no mistakes / 1 mistake / 2 mistakes

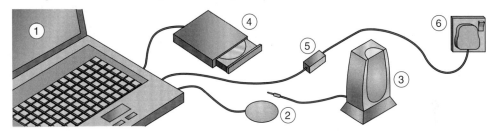

3 Look at the picture in question 2. Correct these sentences.

1   The display is closed.   →   *The display* _____.

2   The power switch is up.   →   _____

3   The mouse is disconnected.   →   _____

4   The speaker is connected.   →   _____

4 🔊 23 Use the words in the box to complete the dialogues. Then listen and check.

| is   isn't   are   aren't   does   doesn't   do   don't |

---

1   *Does* the computer start?
    No, it _____.
    Right. Press the power button again.

2   _____ the power switch down?
    No, it _____.
    OK. Press it down.

3   _____ the loudspeakers connected?
    No, they _____.
    OK. Connect them.

4   _____ the adapter connected?
    Yes, it _____.
    Good.

5   _____ the loudspeakers work?
    No, they _____.
    OK. Connect them and try again.

6   _____ the two LED lights on?
    Yes, they _____.
    Good.

7   _____ the computer start?
    Yes, it _____.
    Good.

8   _____ the loudspeakers work now?
    Yes, they _____.
    Good.

3 User guide

1 Read the Troubleshooting Guide. Underline the correct words.

Travelling with your notebook computer

Close / <u>Open</u> the display.

Press / Turn the power button.

If the display light is low, check / replace the LED for the battery.

If the battery is low, connect / recharge it.

If the battery still doesn't work / start, replace it.

At the end, to turn off your computer, touch / press the power button.

Close / Open the display.

---

2 Make sentences with 'if' from the dialogues.

1 Are the LEDs on?	3 Is the printer light on?	5 Do the speakers work?
No, they aren't.	No, it isn't.	No, they don't.
OK. Check the battery.	OK. Push the 'On' button.	OK. Connect them to the computer.
2 Does the printer work?	4 Are the batteries old?	6 Does it print in black?
No, it doesn't.	Yes, they are.	Yes, it does.
OK. Connect it to the adapter.	OK. Replace them.	OK. Press the button for 'Start Colour'.

1 *If the LEDs aren't on, check the battery.*

2 _____

3 _____

4 _____

5 _____

6 _____

3 Use verbs from the box to complete the Troubleshooting Guide.

check   check   connected   plug   plugged   press   press   shut   shuts   turns   unplug

(1) *Check* that the power cable is (2) *plugged* into the computer and a power outlet.

(3)_____ that the mouse and the keyboard are (4)_____. Unplug the cables and then (5)_____ them in again.

(6)_____ the power button on the back of the computer for a few seconds. This (7)_____ down the computer.

If you cannot (8)_____ down the computer, (9)_____ the power cable from the computer. Wait 30 seconds. Plug it back in. (10)_____. the power button. This (11)_____ the computer on.

4 Word list

NOUNS	NOUNS	VERBS	ADJECTIVES
acceleration	hovercraft	accelerate	closed
adapter	key	check	connected
airboard	laptop	connect	disconnected
battery	LED	contain	fibreglass
body	lever	control	flat
brake	modem	drive	flexible
computer	mouse	force	open
cushion	notebook computer	hold	rubber
diagram	platform	increase	**PREPOSITIONS**
disk	power button	open	above
disk drive	power outlet	press	attached **to**
display	power source	propel	below
engine	purpose	recharge	connected **to**
fan	rider	release	mounted **on**
friction wheel	router	replace	suspended **from**
front	screen	start	**PHRASES**
function	skirt	steer	pull (air) in
handlebar	speaker	stop	push (air) out
hotline	speed	support	suck (air) in
	start button	touch	switch off
	starter motor	turn	switch on
	switch	**ADVERBS**	take out
		backwards	turn off
		downwards	turn on
		forwards	You're welcome.
		upwards	

1 Find 20 nouns that are connected to computers.

_____

_____

2 Replace the underlined words with opposites from the Word list.

1 The rider <u>presses</u> the lever.   →   _The rider releases the lever._

2 The fan <u>pushes</u> air <u>out</u>.   →   _____

3 The rider can go <u>forwards</u>.   →   _____

4 The engine is <u>above</u> the platform.   →

_____

5 The fan <u>starts</u> and the airboard goes <u>upwards</u>.   →

_____

1 Rules and warnings

1 Label the objects with these words.

safety boots   safety goggles   safety gloves   safety helmet

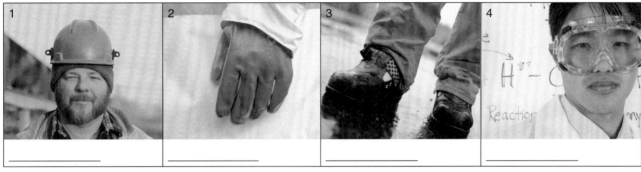

| 1 | 2 | 3 | 4 |

_____   _____   _____   _____

2 Use the words in the box to complete the instructions.

always   do   don't   must   mustn't   never

1   *Don't* smoke in the workshop.
2   _____ use mobile phones in the workshop.
3   You _____ wear safety goggles when you use this machine.
4   You must _____ enter the cold store if you are alone in the factory.
5   _____ not lift heavy weights by hand.
6   You _____ use this machine without the guard.
7   _____ read the manual before you service the machine.
8   _____ touch packets in the cold store without gloves.

3 Complete each sentence with a pair of verbs.

drop / break   lift / hurt   pick / burn   put / melt   touch / get
use / scratch   use / trap

1   Don't *drop* that box. You might *break* the TV inside it.
2   Don't _____ the CD on that hot surface. It could _____.
3   Don't _____ that box without a forklift truck. You might _____ your back.
4   Don't _____ a hook when you lift the car. You might _____ it.
5   Don't _____ up that hot plate. You might _____ your hand.
6   Don't _____ that wire. You could _____ an electric shock.
7   Don't _____ that machine without a guard. You could _____ your hand in it.

2 Safety hazards

1 An inspector is inspecting a factory. Write sentences from his notes.

1 liquid on floor   *There is some liquid on the floor.*

2 hole in the outside door   _____

3 no fire exit   _____

4 broken window   _____

5 cables on a workbench   _____

6 no fire extinguishers in factory   _____

7 2 machine guards missing   _____

8 some damaged warning cones   _____

2 Use phrases from the box with *might* or *could* to complete these warnings.

burn your hand   fall into it   get an electric shock   injure your head
start a fire   trap your hair in it   trip over them

1 Mind that lighted match! (could) *You could start a fire.*

2 Mind that cable! (might) _____

3 Mind those bricks! (could) _____

4 Mind that machine! It doesn't have a guard. (might)

     _____

5 Mind the gap! (could) _____

6 Mind that low beam! (might) _____

7 Mind that circular saw! It's very hot. (could)

     _____

3 Complete the inspector's report about the hazards in a factory. Use each of the words or phrases once.

there was   there were   was   were   two   no   some   the

1 There *were* no fire extinguishers anywhere in the factory.

2 There was _____ food and drink on the workbenches.

3 _____ some boxes of parts on the stairs.

4 _____ guard on one of the machines was broken.

5 _____ some oil on the floor.

6 _____ of the windows were broken.

7 The fire exit _____ locked with a padlock.

8 There was _____ key for the padlock.

3 Investigations

1 ▶ 🌐 **24** Listen to the dialogue. Complete the details on the accident report form.

About the accident	Type of accident (tick)	About the injured person
Date: _____ Time: _____ Location: _____	[ ] injured self [ ] injured somebody else [ ] slipped, tripped or fell [ ] lifted something [ ] dropped something	Name: _____ Job title: _____ Injury: _____ At work: Yes / No (circle)

2 ▶ 🌐 **25** Listen to the questions from question 1. Complete the questions that you hear.

1 First, *where* did the accident *happen*?
2 Was _____ hurt?
3 When did it _____ _____?
4 What's the name of the _____ person?
5 What _____ he _____?
6 What's his _____?
7 Did he injure _____ _____?
8 What _____?

3 Read the newspaper story. Complete it with the words from the box.

away   between   in   in   into   into   of   on   on   on   on   out   with

---

6 FISHERMEN RESCUED

(1) *On* March 19th, there was an accident (2)_____ the North Sea. A cargo ship crashed (3)_____ the fishing boat *Marianna*. The accident happened in the North Sea (4)_____ dense fog, 300 kilometres east (5)_____ Hull. The cargo ship was (6)_____ a journey from Sweden to Portugal (7)_____ a cargo of  2000 tons of wood. The *Marianna* was (8)_____ its way back to Hull, after a four-day fishing trip. There were six fishermen on it. The captain said later: 'The anti-collision system on our boat switched (9)_____ automatically. Suddenly I saw the Swedish cargo ship. The distance (10)_____ us was only 30 metres. I tried to steer our ship (11)_____ from it. But it hit us and our boat sank. We launched our life raft, got (12)_____ it and sent (13)_____ a radio signal for help. We were in our life-raft for four hours.'

---

4 Word list

NOUNS	NOUNS	VERBS	PHRASAL VERBS
altitude	match	coil	look out
aviation	oven	hurt	take care
chemical	padlock	injure	take place
cloud	poison	investigate	**ADJECTIVES**
cone	prohibition	light	bare
distance	safety	lock	careful
drink	shock	mind (your head)	circular
emergency	sign	prohibit	dense
factory	site	service	mandatory
flight	surface	slip	military
food	type	touch	round
gap	warning	trap	single
gear	weight	trip	triangular
glove	**COMPOUND NOUNS**	warn	
guard		wash	
hazard	anti-collision system	wear	
high-voltage	fire exit		
hook	fire extinguisher		
investigation	flight path		
laser	mobile phone		
liquid	near miss		
machine	sea level		

1 Write adjectives to complete these phrases from the unit.

1 *dense* cloud
2 b_____ hand
3 l_____ match
4 n_____ miss
5 c_____ saw
6 e_____ shock
7 h_____-voltage

2 Complete these compound nouns from the unit.

1 *fire* extinguisher
2 s_____ boot
3 s_____ hazard
4 s_____ level
5 f_____ path
6 m_____ phone
7 b_____ site

3 Find ten nouns and compound nouns in the Word list that come from the story on page 78 of the Students' Book. Write them here.

altitude, _____

Section 1

1 Complete the description with the nouns in the box.

| acceleration | body | cushion | engine | fans |
| fibreglass | levers | platform | skirt | |

A hovercraft moves over land and water on a
(1)_____ of air. A powerful
(2)_____ drives four large (3)_____.
They suck the air in and push the air
downwards under the rubber (4)_____. The engine is mounted on a
strong (5)_____. The (6)_____ of the hovercraft is made of foam
covered with (7)_____. Two (8)_____ control the speed of the fans
and the (9)_____ of the hovercraft.

2 Complete the dialogue with a hotline technician. Use the words in the box, some of them more than once.

| can |
| does |
| doesn't |
| have |
| haven't |
| is |
| I've |
| there's |

Technician: Printer Hotline. How (1)_____ I help you?

Customer: I've got a problem with my printer. It
(2)_____ work. (3)_____ switched it on. But
(4)_____ no light on the display.

Technician: OK. First, (5)_____ the printer connected to the AC adapter?

Customer: Yes, it (6)_____.

Technician: Good. And (7)_____ you connected the adapter to the power source?

Customer: Yes, I (8)_____.

Technician: And (9)_____ you turned the switch on?

Customer: Ah, no, I (10)_____. I'll do that now.

Technician: And (11)_____ the printer work now?

Customer: Yes, it (12)_____. Thanks.

3 Look at the flow chart for a personal watercraft. Use the chart to write a Troubleshooting Guide, using sentences with 'If'.

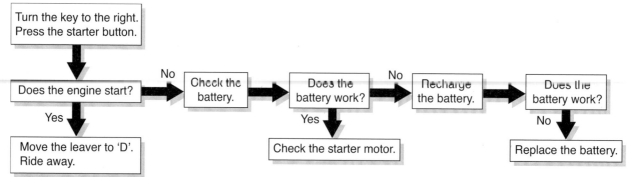

1 *Turn the key to the right. Press* _____.

2 *If the engine starts,* _____.

3 _____

4 _____

5 _____

6 _____

Section 2

1 Write these words for some safety signs in the correct order.

1 use off this after machine Turn → *Turn* _____.

2 shipyard in boots use the safety Always → _____

3 the report to Drivers must office → _____

4 the truck ride Never forklift on → _____

5 use goggles Do without machine not this safety → _____

6 supervisor reverse without You not a must → _____

2 Complete the safety report with the correct form of the verbs in brackets.

Moderna Shipyard

On 31st May, I (1)_____ (inspect) the Moderna shipyard. I (2)_____ (find) a number of safety hazards. There (3)_____ (be) some tools on the ground. There (4)_____ (be) some wire coiled outside the office. Eight of the workers (5)_____ (not have) hard hats. There (6)_____ (be) no safety signs.

In April, there (7)_____ (be) a serious incident in the shipyard. Two ships (8)_____ (be) in the shipyard. A crane (9)_____ (lift) a metal beam from one of the ships into the air. But there (10)_____ (be) no rope attached to the beam. The beam (11)_____ (come) downwards. But it (12)_____ (move) in the air and (13)_____ (hit) a forklift truck. The top of the forklift truck (14)_____ (be) bent. The driver was lucky. Just before the accident, another worker (15)_____ (shout), 'Mind your head!' The driver of the forklift truck (16)_____ (see) the beam and he (17)_____ (not be) hurt.

---

3 Write questions for these answers about the incident in question 2.

1 Q: (Where / incident) *Where did the incident happen?*

A: It happened in the shipyard.

2 Q: (When / take place) _____

A: It took place in April.

3 Q: (take place / on a ship) _____

A: No. It took place near a ship, under a crane.

4 Q: (What / crane / lift) _____

A: It lifted a metal beam from the ship.

5 Q: (rope / attached / to the beam) _____

A: No, there wasn't.

6 Q: (beam / hit / worker) _____

A: No. It hit his forklift truck.

7 Q: (worker / hurt) _____

A: No, he wasn't hurt.

Cause and effect

1 Pistons and valves

1 Complete the description of a flush toilet with words from the diagrams. Some words are used more than once, e.g. tank (4 times).

inlet valve ②
float arm
float ball ②
lever
inlet pipe
tank ④
piston ②
downpipe ②
② handle

Pull the (1) *handle* down. A (2)_____ inside the tank pulls the (3)_____ up. This forces water out of the (4)_____ through the (5)_____.

When you release the (6)_____, the (7)_____ drops back into its place at the bottom of the tank. All the water flows out of the tank through the (8)_____.

There is a float ball inside the tank. It is attached to the end of the float arm. The ball makes the arm rotate vertically. The (9)_____ sinks to the bottom of the empty (10)_____. The (11)_____ rotates and opens the (12)_____. This lets water flow through the inlet pipe into the tank. The water level in the (13)_____ rises. The (14)_____ at the end of the arm rises too.

The tank fills with water. The end of the float arm now presses against the (15)_____ and closes it. This stops water from flowing into the (16)_____ through the (17)_____.

2 Tick the correct forms for the verbs.

	... it do	... it to do	... it from doing
allow		✓	
cause			
let / make			
prevent / stop			

3 Rewrite these sentences to give similar meanings. Replace the verbs in italics with the correct form of the verbs in brackets.

1 The pump *makes* the water flow along the pipes. (cause)
   The pump causes _____

2 The valves *allow* air to enter the tyres. (let)
   _____

3 The valves *don't let* air escape from the tyres. (prevent)
   _____

4 The sun *causes* the solar panel to heat the water. (make)
   _____

5 The cooling system *doesn't allow* the engine to get very hot. (stop)
   _____

6 The closed inlet valve *prevents* the water from flowing out. (not allow)
   _____

2 Switches and relays

1 Complete the description of a circuit breaker with words from the diagrams. Some words are used more than once, e.g. switch (3 times).

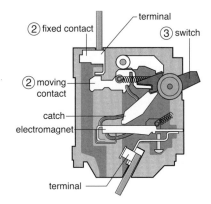

② fixed contact
terminal
③ switch
② moving contact
catch
electromagnet
terminal

A circuit breaker has a (1) *switch* on the outside of the box. You can turn this on or off. When the (2)_____ is up (on), electricity flows into the circuit breaker through the bottom (3)_____. It flows through the (4)_____. It then flows up to the (5)_____ and across to the (6)_____. Then the electricity flows out of the circuit breaker at the top (7)_____.

If the electrical current jumps to a dangerous level, the electromagnet pulls down a (8)_____. This pulls the (9)_____ away from the (10)_____. This breaks the circuit. At the same time, the (11)_____ drops to the 'down' position. The electricity is now shut off.

2 Read the three texts in question 3. Write the titles above the texts.

a) Emergency exit   b) Home security   c) Car security

3 Read the texts and mark the sentences 'True' (T) or 'False' (F). Correct the false parts of the sentences.

---

1 _____

This security system uses a metal ball inside a metal pipe. When the ball remains still, it touches two of the electrical contacts. This completes the electrical circuit. If the ball moves, it breaks the circuit and opens a switch. If somebody makes the car move, the system causes the horn to sound. It makes the car lights go on too. If somebody hits the car a few times, it makes the siren sound.

---

2 _____

The ExitGuard is mounted over the door handle. It has a battery-operated alarm. When somebody breaks open the ExitGuard, this causes the alarm to sound. The ExitGuard allows shops to secure their emergency exits. This stops people from using the exits to enter the shop. But it lets people leave the shop in an emergency.

---

3 _____

If you keep expensive equipment in your workshop, fit a burglar alarm. When the doors and windows of the workshop are shut, the electrical switches are closed. This allows electricity to flow around the electrical circuit. If a burglar forces open a window, this breaks the circuit. This causes the burglar alarm to sound. If your workshop is a long distance away, install the alarm buzzer inside your house. This allows you to hear the alarm when you are inside your house.

---

1 <u>Two</u> of the systems cause an alarm to sound. *F   Three*

2 The car security system works when somebody moves the car.

3 The ExitGuard works when somebody touches the door.

4 If there is a fire in a store, people can break open the ExitGuard.

5 The burglar alarm works only on the windows.

6 If the electrical circuit is broken, the burglar alarm will sound.

3 Rotors and turbines

1 Hidden word puzzle. Write the words for the pictures.

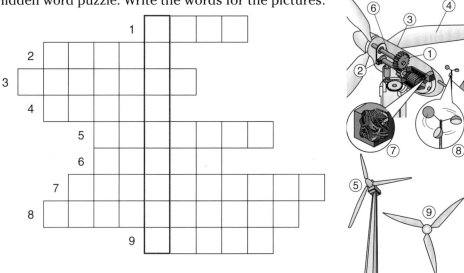

2 ▶ 🎵 **26** Write the ten words from question 1 next to their explanations. Then listen and repeat.

1 this produces electricity from the wind   *turbine*

2 this measures the speed of the wind _____

3 this switches the wind turbine on and off _____

4 this slows down the rotating shaft _____

5 two of these make the high-speed shaft rotate at 1200 rpm

_____

6 the wind blows on three of these _____

7 this produces AC electricity _____

8 this transmits rotation to the generator _____

9 this is a strong rigid container _____

10 this consists of three blades and a hub _____

3 ▶ 🎵 **27** Listen to the interview with a company technician about a wind farm (a group of wind turbines). Circle the correct information.

1 In which state of the USA is the wind farm?

a) New Mexico   b) Texas   c) Nebraska

2 When did the first part of the wind farm open?

a) 2005   b) 2006   c) 2007

3 How many wind turbines are there on this wind farm?

a) 130   b) 291   c) 421

4 Is this the largest wind farm in the world today?

a) the largest   b) the second largest   c) the third largest

5 How many wind farms does the company have?

a) 38   b) 48   c) 58

4 Word list

PISTONS & VALVES	SWITCHES & RELAYS	ROTORS & TURBINES	
NOUNS	**NOUNS**	**NOUNS**	**ADJECTIVES**
guard	bell	anemometer	powerful
high pressure	burglar	blade	simple
low pressure	buzzer	brake	**ADVERBS**
inlet valve	circuit	controller	automatically
outlet valve	conductor	data	**VERBS**
overflow pipe	contact	gear	download
piston	earth	generator	click
piston pump	electromagnet	high-speed shaft	**PREPOSITIONS**
shaft	magnet	housing	next to
spring	relay switch	hub	**NOUNS (NOISES)**
trigger	strip	low-speed shaft	alarm bell
VERBS	switch	rotor	beep
allow	wire	tower	buzzer
cause	**VERBS**	wind turbine	click
contract	buzz	**VERBS**	dial tone
decrease	sound	blow	door bell
expand	spring	contain	horn
explode		transmit	siren
force			
increase			
let			
prevent			
pump			
spread			

1 Find opposites in the Word list for these words.

1 allow _____
2 contract _____
3 increase _____
4 suck _____
5 receive _____
6 inlet _____
7 high pressure _____
8 low-speed _____

2 Find noises for these things.

1 the *beep* of an answerphone
2 the _____ of a car
3 the _____ for a fire
4 the _____ of a mouse
5 the _____ of a telephone
6 the _____ of a police car
7 the _____ on a door (2 choices)

1 Data

1 Circle the names of 15 words from the text on Students' Book page 90. Some words are plurals. They go vertically from top to bottom, and sideways from left to right. No words go diagonally.

A	S	P	E	E	D	E	B	I	C	Q	O	B	F	D
E	U	D	J	D	L	M	O	T	O	R	S	U	H	K
H	S	F	M	K	A	Q	G	I	Z	D	L	C	E	G
A	P	W	A	I	X	A	N	T	E	N	N	A	S	M
Q	E	B	S	Z	O	F	Q	A	C	Y	F	M	W	U
I	N	S	T	R	U	M	E	N	T	S	Q	E	H	D
R	S	G	F	V	G	N	V	I	W	K	L	R	Z	A
D	I	A	M	E	T	E	R	U	D	H	V	A	R	F
X	O	Q	Z	W	K	Y	H	M	V	G	B	S	X	G
E	N	B	X	H	V	I	X	O	L	R	O	B	E	R
K	F	Y	K	E	L	R	K	Z	Q	O	D	F	W	H
G	L	A	S	E	R	G	U	N	G	B	Y	Y	U	R
B	W	Q	D	L	Y	A	Z	D	Z	O	W	I	G	Q
T	O	O	L	S	K	R	G	E	H	T	F	L	K	O

2 Read about the underwater robot *Jason*. Cross out the incorrect words.

Jason

Jason is an underwater science laboratory. It weighs, (1) *in/on* the surface, a little (2) *over/near* 3600 kg. It can operate (3) *in/at* a maximum depth of 6500 metres. There are some cameras and lights mounted (4) *on/over* its body. When it is (5) *about/near* the sea bed, the cameras look (6) *around/over*.

Jason has two robot arms attached (7) *on/to* the front of its frame. There are special tools (8) *from/at* the end of each robot arm. Some of the tools collect water samples. Others collect rocks (9) *over/from* the sea floor. A small pump sucks (10) *on/in* living things. An instrument measures the temperature of the water. Jason 1 started work in 1988 and worked (11) *up to/over* 2001. The new *Jason 2* has made 183 dives. It has spent (12) *near/at least* 3249 hours at the bottom of the sea.

3 Write questions for these answers about *Jason*. Use the information in question 2.

1 Q: What is _____?

    A: It's called *Jason*.

2 Q: What _____?

    A: A little over 3600 kilos.

3   Q: Where _____?

     A: They're attached to the front of the frame.

4   Q: How _____?

     A: A small pump sucks them in.

5   Q: Where _____?

     A: They're at the end of each robot arm.

6   Q: How many _____?

     A: At least 183.

2 Instructions

1   A controller is training a mobile crane driver. Match the phrases in the two boxes.

1 Press	a) the 'Start' button.
2 Press	b) forwards.
3 Release	c) to the left.
4 Push the joystick	d) the power switch. *1*
5 Turn the wheel	e) the hand-brake.
6 Move forwards	f) 45° to the left.
7 Press	g) about 50 metres.
8 Rotate the arm	h) backwards.
9 Pull the joystick	i) backwards 10 metres.
10 Reverse	j) the brake pedal.

2   Make sentences with verbs and phrases from the box.

> flies   goes   up and down   forwards and backwards
> into space   over rocks and holes

1   A car *goes forwards and backwards.*

2   A helicopter _____ _____.

3   A motorboat _____ _____.

4   A plane _____ _____.

5   A rover _____ _____.

6   A shuttle _____ _____.

7   A truck _____ _____.

3   Re-read paragraph 1 of the text from Section 1 Data, question 2. The ship's crane is now lifting *Jason* from the sea floor. Put the verbs in brackets into the present continuous.

A: Now, lift *Jason* up to the surface. Pull in the wire.

B: It (1 not move) *isn't moving*. *Jason* (2 not come) _____ up. I think it's stuck to some rocks.

A: Move the arm of the crane to the left. Now raise the arm of the crane.

B: I (3 bring) _____ it up.

A: What (4 happen) _____ now? (5 move) _____ the craft _____?

B: Not yet. It's stuck.

A: Move the arm to the right. Bring the arm up suddenly. Now pull in the wire.

B: I (6 pull) _____ it in now. Oh no!

A: What (7 happen) _____?

B: The wire (8 come) _____ in very fast now. I think the wire is broken. And *Jason* (9 sit) _____ on the sea floor.

3 Progress

1 ▶ 🔊 28 Change these sentences, using verbs from the box. Give sentences 1–5 the opposite meaning. Then listen, check and repeat.

    assemble   attach   connect   loosen   replace   take   tighten

1 Bring the large wrench from the workshop.

    1 *Take the large wrench to the workshop.*

2 Loosen the nuts on the supply pump.

    2 _____

3 Remove the pump from the supply pipe.

    3 _____

4 Dismantle the water pump.

    4 _____

5 Disconnect the valve from the pump.

    5 _____

    6 *Replace* the valve.

2 ▶ 🔊 29 The manager of an F1 racing team is talking to the engineer. Mark the jobs on the chart with a [✗] or a [✓]. Write the days/dates for finishing the jobs.

Task	Y/N?	Date for finishing
1 Remove nose cone.	✓	May 17th
2 Take photo of nose cone.		
3 Inspect fuel tank.		
4 Replace fuel pipe.		
5 Attach cables to foot pedals.		
6 Install new valves.		
7 Lubricate the gears.		
8 Test the car.		

3 Check your answers for question 2 in the Answer key. Correct them if necessary. Write sentences about the eight jobs below.

1 (they) *They've removed the nose cone.*

2 (he) *He hasn't* _____. *He'll* _____

3 (they) _____

4 (they) _____

5 (he) _____

6 (he) _____

7 (they) _____

8 (they) _____

4 Word list

NOUNS	NOUNS	VERBS	VERBS
astronaut	oxygen	analyse	remain
camera	photograph	assemble	remove
control centre	powder	check	replace
diameter	progress	collect	respond
distance	range	confirm	roll
equipment	robot	dig	support
instrument	rover	dismantle	train
laser beam	simulation	fire	**ADJECTIVES**
laser gun	surface	include	average
mass	suspension	inspect	daily
mast	system	install	mobile
million	ventilation	lubricate	scientific
obstacle	waste	operate	**ADVERBS**
		orbit	approximately
		prepare	over
		range	less than
			more than
			under

1 Find opposites in the Word list. Write them here.

1 assemble _____

2 install _____

3 leave _____

4 exclude _____

5 more than _____

6 over _____

2 Combine two nouns, one from each box. Write them below.

laser	range
robot	system
six-wheel	tank
science	beam
suspension	arm
temperature	drive
waste	laboratory

laser beam, _____

_____

Section 1

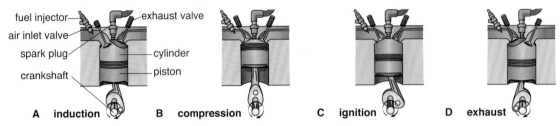

| A | induction | B | compression | C | ignition | D | exhaust |

1 Read about the four-stroke petrol engine. Correct the sentences, using the verbs in the box.

allow   cause   let   make   prevent   stop

A. The inlet valve opens. The round metal piston moves downwards. It stops the pressure inside the cylinder from falling.

(1)   *It makes the pressure inside the cylinder fall.*

This doesn't allow a mixture of petrol and air to enter the cylinder.

(2) _____

B.   The inlet valve closes. This lets the fuel mixture escape.

(3) _____

The piston moves upwards. This prevents the pressure in the cylinder from rising.

(4) _____

C.   The spark plug lights the fuel and doesn't cause it to explode.

(5) _____

This forces the piston downwards on its power stroke.

D.   The outlet valve opens. The piston moves upwards. This stops the burnt fuel from escaping.

(6) _____

2 Complete the description of an electricity generating station in France. Use words from the box.

blades   cables   dam   electricity   gates   generator   shaft   turbine

The (1) *dam* across the River Rance in France was finished in 1966. Water flows from the river into the sea through the (2)_____ in the dam. Later, it flows back into the river from the sea. The water flows past the (3)_____ of a (4)_____ and makes it rotate. A (5)_____ connects the turbine to a (6)_____. The rotation of the generator produces (7)_____. The electricity leaves the power station through high-voltage (8)_____.

Section 2

1 Write sentences about an overland rover. Use the information from the specifications chart. Change the abbreviations to words.

1 Height	202 cm	5 Wheels	steel alloy
2 Length	365 cm	6 Wheel diameter	52 cm
3 Weight	3050 kg	7 Max speed	155 kph
4 Drive	4-wheel	8 Max/Min temperature range	−40 °C min to +55 °C

1 *The rover is 202 centimetres high.*

2 *It has* _____.

3 _____

4 _____

5 _____

6 _____

7 _____

8 _____

2 Look at the progress chart for May 4th and complete the dialogue.

Task	Yes/No?	Date for finishing
1 Collect rock samples from the rover	Y	May 4
2 Repair solar panel	N	tomorrow
3 Connect cables to solar panels	N	tomorrow
4 Replace damaged mast	Y	May 2nd
5 Assemble new robot arm	Y	May 1st
6 Replace bent wheel	N	in progress
7 Service the brake system	N	May 7th

A: Now, it's May 4th today. (1) *Have you collected* the rock samples from the rover?

B: Yes. We (2) *collected* them today.

A: Right. What about the solar panel? (3) *Have you repaired it yet?*

B: No, we (4)_____. We'll (5)_____ tomorrow.

A: Have you (6)_____ the cables to the solar panels?

B: No, not yet. (7)_____ tomorrow.

A: Right. What about the damaged mast? Have (8)_____?

B: Yes, we have. We (9)_____ May 2nd.

A: Right. What about the new robot arm? (10)_____?

B: Yes, we (11)_____ May 1st.

A: Have you (12)_____ the bent wheel yet?

B: No, we're still (13)_____ that.

A: What about the brake system? (14)_____?

B: No, we haven't. We (15)_____ May 7th.